CMM:
A Beginners
Understanding

I0474211

Michael Mino

DEDICATION

To a friend that no longer is there...thanks for 10 years.

CONTENTS

ACKNOWLEDGMENTS

I would first like to present the purpose of this book to you. By the conclusion of reading this book, you WILL NOT be able to go in and immediately start programming a Coordinate Measuring Machine, or CMM for short. This book is intended to be used as a beginners guide to understanding what a CMM does, how it works, and to understand some key pieces of information that is like GOLD when first starting out working with a CMM. This book is not to teach you how to write a program. This would be a very tough task, due to the many different programs that are used to write programs. This book is to maybe help you understand the CMM from truly a beginners look. First, let me give you a little back story on why I am writing this book, as well as how I began to enter the world of writing programs and running a CMM.

CHAPTER 1

STARTING OUT

I took a job that was of higher pay at my workplace, which happened to be writing and running programs on a CMM. When first using the CMM, I had a quite a bit of trouble grasping how a machine like this could measure in a three-dimensional view. I did some research that night on the internet on the basics of running a CMM. What I found was not helpful at all. It was if they were all talking in a different, foreign language. I thought that there had to be a better way, and it turns out, there was. I had a heck of a teacher that helped me pick up on the CMM very quickly, which was a co-worker. He explained things in normal terms, not the scientific, expert way. This was very helpful, and I immediately thought that if there was a book of what this guy was teaching me, I would pick it up in a second. It was so helpful that with his foundation of teachings, I was able to wrap my head around the CMM concept and pick it up. That is the mission I am presented with today, with you, the reader. The purpose of this book is to lay the foundation of what a Coordinate Measuring Machine does. This guide is for beginners only, so if you have experience in the CMM world, I suggest you move on a higher level of literature.

So lets get started to try and wrap out heads around the concept of how a CMM measures certain subjects.

CHAPTER 2

CHECKLIST

There are many things that must be done before actually running a Coordinate Measuring Machine. Lets take a look:

1. Wipe down and clean everything you will be working with. It is EXTREMELY important that your CMM table is level, free of dust and debris, and sturdy. This is key, because you are dealing is such small detail, that any rocking or movement of any kind will throw off your dimensions. Make sure to also keep the subject you are measuring stable as well. Be sure to keep hands clean as well. Wipe the probe off as well as table and subject. Believe it or not, this is very crucial. Any dirt or debris can throw off your measurements.

2. Make sure the CMM is within calibration. Your specific CMM comes with recommendations of how often that particular machine needs to be re-calibrated. This is done with a calibration tool, and should be done very often. Dependent on location of the CMM, temperature, and many other factors, the calibration should be done on a set schedule and very often. Every machine also is receptive to the temperature that it is being held. Simple temperature changes could effect measurements and machines.

3. It is a good idea to check for any updates for your software that you will be using to program with. Keeping the newest version of software ensures minimal bugs and crashes within that software will occur. This will ensure proper compatibility with the actual Coordinate Measuring Machine and the software that your are using.

4. Be careful operating the CMM. I have seen many people run the probe into the table or into the subject being measured because of time constraints or not paying attention. These are very costly mistakes, and can set you back on schedules. Also, be aware of your surroundings. Many machines have arms that stick out and can hit unsuspecting bystanders. This can be costly as well. Make sure the surrounding people know that you are measuring the subject as well. This will ensure that no one will tamper with the subject while you are away or writing your program.

5. Make sure to have a copy of the drawing or blueprint of the actual subject that your are writing your program on. This comes in very handy, and allows you to make certain programming notes on it, without messing up any master copies.

CHAPTER 3
GEARING UP

The first thing to know about running a CMM is that it measures on a host of planes. The most common is the X,Y, and Z planes. Typically, if you are looking at a subject, the X plane would run right to left, left to right. The Y plane would be back and fourth, away from you, then to you. The Z would be up and down. This is the most standard planes that I will be talking about in the beginners guide. There are more, but this book is to keep it simple at first. It is possible to switch these planes, for example, switch the X and Y, but this will come later. Again, basics.

Try to imagine the Z plane being how tall you are. The X is how wide, and the Y how fat. This is a funny example to work with, but its the most personable. Again, these planes can be changed within the program, but its important to get the standards down.

I am assuming that since you are reading this, that you have seen a CMM running. Note how careful the operator is watching the program which is running. Any problems with the program that is written, or incorrect subject being measured could cause a crash. This can be dangerous to the equipment, subject, and surrounds, so always keep a close eye on the machine.

So, you have your machine and subject clean? Is it in proper calibration? Is the software updated with the latest version? No outside interference? Then lets go!

CHAPTER 4
RED LIGHT

On a CMM, you will be taking what they call "hits" with the probe. Think of these hits as imaginary dots. These dots are the basis of what you will be measuring with. For now, we will call these hits "points". With these points, you will need to create an alignment. This is done by creating a plane, a line, and a point. Know that phrase well, because almost every program need this. Note to plan these out in advance. These are the actual o's that you will be measuring from. Think of them as where you hold the start of the tape measure.

A plane is created with a series of 3 or more hits on the same surface. Most of the time, and depending on which software you are programming on, it will detect that the points you have consecutively taken were on the same plane, therefore labeling it "Plane 1" This can be on the X, Y, or Z plane. I generally keep the plane 1 on the Z plane, but this is by programmers choice, as well as dependent on what you subject you are measuring. It seems that most software accepts this much better, and that is key when you just starting out.

A line is created by using 2 or more points. You can construct a line by taking 2 points that you want your line to be made of. So for example, if you took a point on one edge of a desk, then the other side of the edge, you could construct a line between these two points. Then this line would be your zero for that plane, and you would be measuring on the Y.

We already know what a point is, so just remember that when selecting this point, it will be the zero for whatever plane you plan to use it on. Now you are ready to set up an alignment.

When you do, just be sure to remember which planes you were using for which items created, and you will do fine.

CHAPTER 5
YELLOW LIGHT

Upon setting up your initial hits, you must have a plan ready. A quick look at the blueprint or drawing of the subject is extremely helpful. Go through and understand where you need your starting points at. If a lot of measurements are taken off of a certain detail or spot, then this is ideal for a potential alignment candidate.

Plan out your points around the drawing or blueprint of the subject. Think of this as a trail that your future program is going to run. It would not be wise to have it take a point on the subject all the way to the right, the left, then back to the right. Try to make points in consecutive order and have the least travel time as possible. Depending on how the software you use, and the way you will program, this will help you tremendously when you first start writing your program. Note any measurements that you need, and ones you do not. It is a good idea to get a copy of the drawing or blueprint that you can write on to make notes of where you want your points, measurements, and alignments. This comes in very hand as well, because it helps you to remember certain measurements that are critical to the program itself. You can write all of the future points on this as well as angle lines, planes and anything else you might need to make notes of while programming your subject.

CHAPTER 6
GREEN LIGHT

So with this all set up now, you can now measure the location of certain points, and they will measure back to the alignment given. For instance, if you set the floor as your Z plane zero, then all you would need to do is take a point on the exact top of your head. We will label this "Point 5". Now you can select the location of Point 5, and it would automatically give you the distance between plane 1 and point 5, which would be how tall you are. Granted, this example is not a good one, given so many changing variables, but remember, this is a beginners guide. Any example that I can give that helps you understand a little bit better is priceless.

So with the given example, if the floor was Z plane Zero, you could take hits all the way down your body, from your head (point 5), to your nose (point 6), to your chin (point 7), to your belly button (point 8) and to your knees (point 9). Then in your program, you could measure the location of all these points, from point 5 to point 9. If you wanted a specific distance between points, you can do this as well. If you needed the measurement between your nose to your knees, you could just summon up the distance between point 6 and point 9, on the Z plane. The difference between "distance between", and "location", is that the location of a point or feature is measured back to the alignment setup, in most cases.

Now to get into the more 3 dimensional area, if you were to extend your arm out, and then take a point off of the tip of your extend finger of your hand, then this could become Point 10. By doing this, we have now added another plane into the measuring picture. You have a Z plane, and a X plane.

CHAPTER 7
MULTIPLE PLANES

Now that we have entered the third dimensional side of things, lets consider the fact that we can now measure the points in the previous chapter in two ways. In most programs, you can opt to measure the dimension in 2D or 3D. This helps you understand which dimension you are looking for, versus what you are actually getting in your program. You could measure the distance between point 10 and the Z plane zero in 2D or 3D. Again, check with your current software help section to decide if this option is available.

Three dimensions can become very tricky to picture in your mind, but with a little practice, you will grasp it. Try to picture the subject you are measuring in your mind. Now rotate it back and forth, then side to side. Now roll it. Picture the points that you have taken as red dots, and this might help you understand the three dimensional views better when adding and inserting your dimensions that are necessary.

CHAPTER 8
ANGLES

Angles can also be measured on most CMMs. You could take a two points off the front edge of your desk, then create a line within the program. Now take two points off of the side edge of your desk. Create another line. We will call these lines 1 and 2. Now, within whatever programming software that you are using, you can measure the angle between these two lines. This would normally be a nominal of 90 degrees, because it is a 90 degree angle. Depending on how crucial the angle is on the subject would determine if the subject would pass its inspection or not.

There is so much more to measuring angles, but I don't want to go very much in depth with this section. The example above was a very elementary example, as well as all the examples in this book. For more information on angles, consult your software help center that your are going to be programming with. There is plenty of information within these help centers of your software, that they can give you further help with understanding angles, as well as software specific information. Use this tool as much as you can while just starting out.

CHAPTER 9
FINAL WORD

Again, everyone's programming software is different. It is important to read the tutorials and help centers of your software that you are using. Some of this information that I have mentioned may not apply to your current software, or may even be outdated. Many programs have a graphical interface which allows the programmer to actually see in wire frame or 3D models of where the points were taken. I'm sure in the future, we will see much advancement in the CMM programming world. Whatever the case, CMM programming is a great field to get into today, and a great thing to learn. The field is getting in high demand, which is great for programmers like yourself.

Again, I cannot stress enough that this book is just for beginners. I know that I haven't even touched on many steps of programming. The intention of this book is to give an explanation of what a CMM does, and to maybe have a better grasp going forward to learn how to program. You now know a little bit about planes, points, measurements, angles, and the steps to getting ready to learn how to program. I personally want to thank you for reading this book, and wish you the best of luck in a great, promising field.